DECISION MAKING AND ARTIFICIAL INTELLIGENCE IN PUBLIC ELECTRICITY MANAGEMENT

1ª edition

Fabricio Quadros Borges

São Paulo - SP
2024

BORGES, Fabricio Quadros

Decision Making and Artificial Intelligence in Public
Electricity Management / Fabricio Quadros Borges.
São Paulo, SP: AlphaGraphics. Brazil, 2024.
41p.; 29cm.

ISBN: **978-65-00-98918-2**

SUMÁRIO

PRESENTATION

The analyzes carried out by artificial intelligence must start from a complete and integrated data structure, which is classified and grouped with the intention of synergistically producing mental and predictive captures. In this perspective, the objective of this study is to analyze the possibility of contribution of artificial intelligence in guiding decision-making in the public planning of sustainable electrical matrices. The methodological procedures of this investigation, built a structure of analysis of electricity sources, based on the economic, social, environmental and technological dimensions; as well as a sectoral analysis structure of energy sustainability indicators, supported by linear correlations of an economic, social, environmental and political nature.

The planning of electrical matrices, according to the inferences of this investigation, can use artificial intelligence as a strategic guide for decisions, as long as they are based on analysis structures focused on the strategic use of electricity sources and the use of sectoral and multidimensional indicators. This investigation constitutes an original contribution insofar as it discusses the possibilities of connections between artificial intelligence and the construction of electrical matrices, from the perspective of improving the decision-making process in Brazilian public planning.

The discussion about these connections helps to raise subsidies for machine learning to process and develop methodologies, based on algorithms, that automate the construction of decision analysis models in the planning and sustainable construction of the use of electricity.

INTRODUCTION

Studies on A.I. emerged in the 1950s, through scientists Herbert Simon and Allen Newell, who created the first artificial intelligence laboratory at Carnegie Mellon University (Santos, 2021). The advances in Artificial Intelligence (A.I.) bring important reflections on how this field of knowledge will affect human routine. The Web Summit 2018, one of the largest technology and innovation conferences in the world, points to a solid and irreversible path towards this technology. The AI branch seeks, through computational symbols, to elaborate mechanisms that simulate the human capacity to be intelligent, which already demonstrates effects in several areas of knowledge; the advances of these studies represent improvements in the application of available resources to the extent that they can strategically guide the decision-making process in organizations (Colson, 2019; Santos, 2021).

In view of this panorama, there is a hope that A.I. it could also help public planning in the electricity sector, insofar as it could support public managers to design more sustainable electrical systems through the optimization of resources, help motivate individual or organizational consumption patterns. It is envisaged that this possibility of contribution is based on the use of analytical models capable of accurately examining the ideal combinations of the use of energy sources in alignment with the specificities of each region and effectively favoring the improvement of the standard of living of the populations. Energy generation sources have advantages and disadvantages that need to be precisely identified and examined from the normative framework of sustainable development.

The energy issue becomes increasingly important in the planning agenda of both developed and emerging countries (Campos *et al.*, 2017). In this perspective, the process of economic expansion of a country is linked to an increase in the supply of energy generated by investments applied in the electricity sector and, therefore, increased consumption (Reis *et al.* 2012; Narayan *et al.*, 2017). The role of electricity through the electricity sector is of fundamental importance within a nation, as it moves all sectors of economic activity within society (Saidi *et al.*, 2017). In view of this, this input has been treated as a good of a strategic nature that involves economic, social, environmental, political and technological dimensions (EIA, 2018).

The conditions of availability of electric energy in quantity, quality and costs determine the ability of societies to ensure a certain standard of living through targeted investments, hence the peremptory need to improve the decision-making process along with the construction of electrical matrices, through of machine learning. Machine learning, a branch of AI, is a methodology for examining information that automates the creation of analysis models, supported by the idea that systems can learn from data and identify patterns (Colson, 2019; Desordi & Bona, 2020). In reality, it is human beings who make decisions, systems only add value to information by transforming it into knowledge that may or may not be used by managers.

Thus, decision-making in environments with multiple uncertain and imprecise information, and the extraction of knowledge from varied and complex databases for the benefit of society, could have its risks severely reduced. In this context, this article asks: how can machine learning guide the public management of electricity in Brazil in a sustainable way? The implementation of A.I. in public planning of energy matrices is an inexorable step for governmental organizations in the Brazilian electricity sector to transform their processes, prioritizing effectiveness and transparency.

Desordi and Bona (2020), developed investigations that analyzed how the use of A.I. can contribute to the realization of the principle of efficiency in Public Administration. Figueiredo and Cabral (2020), especially investigated the insertion of A.I. in the activities carried out by the Public Administration, observing the principles of good administration and the implementation of fundamental rights. Research carried out by Marques (2020) developed an approach focused on the limits and possibilities of using artificial intelligence in the context of public administration. Araujo et al. (2020) carried out an investigation that particularly addressed the impact of administrative decisions made based on algorithms from large databases in the scope of Public Administration. But it was through Valle's investigations (2020), that several conceptualizations on artificial intelligence modeling, especially machine learning, and the functions it can play in public organizations became available.

What is known about the subject is that AI can strategically contribute to the field of Public Administration (Colson, 2019; Desordi & Bona, 2020; Marques, 2020; Araújo *et al.*, 2020 & Valle, 2020). What is not known is how AI can contribute to the public planning of sustainable electricity matrices, in order to consider the potential and borderline specificities of the sources used, as well as, in order to consider significant relationships between electric energy variables and development variables , without neglecting the interactionist sociological aspects. The notion of intelligence used ignores the fact central to interactionist sociology that there is interaction with other human beings in social contexts (Dreyfus & Dreyfus, 1986). This article fits into this context insofar as it seeks to meet these demands through an analysis of the possibility of artificial intelligence contributing to guiding decision-making in public planning of sustainable electricity matrices, in order to consider the specificities of sources of electricity generation, the strategic relationships between energy and development and the interactionist contribution of sociology.

The study brings as learning a contribution in a practical scope, by providing opportunities for a debate that seeks to discuss possibilities of connections between artificial intelligence and the construction of electrical matrices, from the perspective of improving the decision-making process in Brazilian public planning. Specifically, the investigation facilitates this debate through the elaboration of structures for the analysis of electricity sources, based on economic, social, environmental and technological dimensions; as well as, through a structure of sectoral analysis of energy sustainability indicators, supported by linear correlations of economic, social, environmental and political nature. This discussion is relevant insofar as the preliminary elaboration of these multiple structures of dimensional analysis comprises a condition for the precise survey of subsidies for machine learning to process and elaborate methodologies, based on algorithms, that automate the construction of decision-making models in public energy management in Brazil.

MISSION OF PUBLIC ENERGY MANAGEMENT

The public energy management environment is developed through public policies in the electricity sector, which generally aim to demonstrate that the investments aim at economic growth and the improvement of the population's living conditions. In this process, strategic aspects are verified, from the choice of sources of electricity generation to the reflexes of the use of this energy with the different sectors of the economy of a country. The basis of the debate on public management has raised several questions that are important for the analysis of the managers' ability to achieve qualitative results in the uses of public resources applied to the territory; among them, the influence of group ideologies that interfere with more decision-making power, through correlations of forces along various branches, such as energy (Mafra & Silva, 2004).

The development refers both to the physical basis of the growth process, aiming at maintaining the stock of natural resources incorporating both productive activities and the carrying capacity of ecosystems. The relationship between the terms seems to be established through sustainability, where the development paradigm seeks to respond to the dilemmas established by current modernity. Studies on the processes of globalization seek to reinterpret contemporary globalization from the idea of the need to pay attention to the social, political and cultural dimensions, in addition to the economic, which has recognized importance (Santos, 2013).

A discussion on the issue of rationality that questions sociology's competences regarding this analysis comprises the focus of study. According to this research, which does not clearly work with the development category, the author seems to identify with the idea of development based on a rationality (Habermas, 1987). Studies claim that the recognition that speeches on sustainability have reached the center of international environmental policy, proposes a more detailed examination of the current political and intellectual agenda (Redclift, 2003). This logic supports the thought that the idea of sustainability is still useful, but that it should not be associated solely with external nature. He mentions modernity only when he uses Fairhead and Leach who claim that modernity has been progressively destroying the forest, by transforming cultures previously favorable to these forests. The author does not work with a concept of development itself, but constantly relates it to the term Sustainability, which he values for social, economic and environmental balance (Redclift, 2003).

Studies address the economic and developmental assumptions that inform the notion of sustainable development, through a discussion of its consequences culminating in the question: who sustains the development of whom? (Banerjee, 2003).

These studies do not clearly address the category of modernity, however, it moves in the direction of defining development when it notes that sustainable development describes a process of economic growth that does not cause environmental destruction. simultaneously, maximize economic profits and environmental well-being. In view of this theoretical review of interpretations, it appears that the definitions and relations between modernity and development are far from being exhausted and limited, however, we can clearly understand consensus towards the idea that modernity would be characterized , in fact, because it is dominated by the idea of the history of thought as a progressive illumination, which develops based on the appropriation and the fuller reappropriation of the 'fundamentals', which are often also thought of as the origins, that the theoretical and practical revolutions of Western history are presented and legitimized most of the time as recoveries, rebirths, returns. It is from the notion of overcoming that modernity legitimizes development, which in turn represents a progressive enlightenment of thought, which reappropriates and builds its own foundation and origin again.

This development, inclusive and broadening of well-being, would promote the transition to another social structure. Next, based on this review of the literature on modernity and development, an attempt is made to feed the debate through an approach to development as a process of structural transformation of society in an attempt to better understand the realities in public actions in the electricity sector in Brazil Amazon. Development as a process of structural transformation of agrarian (traditional) societies into industrial (modern) societies represented the great theme of political economy. The theoretical discourse of the authors who study development analyzes the environment of the strategies, that is, measures to be adopted for a balanced and self-sustained economic growth in a given society.

The discussion about development must be guided by the systematization of the production of knowledge of the founding themes of the current debate on the theme in order to contribute to the reflection and interpretation of society's dilemmas, as well as to better understand the strategies and political actions with which different social actors act and intervene in the solution of problems related to development. More specifically, it should seek, based on the main theoretical matrices in the environment of the social sciences and economics, to subsidize the understanding of the strategic participation of the electric energy sector in the process of socioeconomic development in the Amazon.

HISTORICAL EVOLUTION OF THE DEVELOPMENT PROCESS

Studies in the 1960s analyzed the historical evolution of developed countries and identified five stages of development (Rostow, 1961): traditional society; prerequisites for grubbing up; yank; self-sustaining growth and age of mass consumption. Traditional society, in general, is predominantly agrarian, with little technology and low per capita income. In the second stage, the preconditions for the take-off are created, based on important economic and non-economic changes. There is an increase in the rate of capital accumulation, in relation to the rate of demographic growth, and an improvement in the degree of qualification of the skilled labor for specialized production on a large scale.

The crucial period is the hitch. At this stage, the process of continuous growth is institutionalized in society. This is because, in the second stage, there is still some resistance, since society is still characterized by traditional productive attitudes and techniques. The fourth stage, that of the march to maturity, takes about 40 years. In its course, modern technology extends from the leading sectors, which propelled the take-off, to other sectors (Rostow, 1961).

The economy demonstrates that it has the technological and entrepreneurial ability to produce anything it decides to produce. Finally, the economy reaches the fifth stage, the era of high mass consumption, when the leading sectors turn to the production of high-tech consumer durables and services. In this phase, income rose to levels where the main consumption goals of workers are no longer basic food and housing, but cars, microcomputers.

There are some criticisms of the theory formulated by Rostow. It would be treated more of an empirical analysis, from the observation of what happened with the developed countries, than a scientific analysis. Many historians do not see a clear distinction between the second and the third stage. Still, Rostow seems to imply that industrial evolution can only happen after the improvement of agricultural productivity, and that they do not occur simultaneously. In any case, the essence of Rostow's so-called Step Theory illustrates the fact that economic development is a process that must advance in a certain sequence of clearly defined steps.

Studies have indicated that societies change in different trends, moving from traditional to modern (Parsons, 1964); thus, these studies present a contribution to restructure the extension of evolutionary thinking in sociology. The author presents three stages of development: primitive, advanced primitive and modern. In the primitive, one finds technology, kinship, communication and religion as basic mechanisms. In the advanced primitive, Parsons cites stratification and legitimation. And finally, in the modern, there is the bureaucratic organization, money and the market, the universal legal system and the democratic association. It is observed that the conception of Evolutionary Universes, developed by Parsons, still provides valuable insight in the interpretation of the complexity and diversity of the historical evolution of many societies.

The author was not convinced with evidence that social evolution follows an organic evolution. When discussing the issue of development x underdevelopment as two sides of a single global process, studies seek to clarify some controversial points about the conditions, possibilities and forms of economic development in countries that maintain relations of dependency with the hegemonic poles of the capitalist system (Cardoso, 1993).

This research warns of the need to consider structural and historical specificities when talking about Latin America and presents three stages of the development process: in the first there is the substitution of imports, then the production of capital goods and in the third, the income redistribution. After the first two stages, in the 1960s, there was a period of relative stagnation in Brazil, thus evidencing that the impression that the interpretative scheme and the predictions formulated in the light of purely economic factors were not sufficient for the later course of events. It would not be enough, replacing the "economic" interpretation of development with a sociological analysis, but integrating them.

Research in the 1990s shows the need for an integrated analysis that provides elements to provide a broader and more diversified answer to general questions about the possibilities of development or stagnation in Latin American countries, and to answer decisive questions about its meaning and its political and social conditions (Cardoso, 1993).

As for underdevelopment, using structuralist reasoning, studies observe that it comprises an autonomous historical process, not constituting a necessary stage for the formation of capitalist economies (FURTADO, 1994).

According to these studies, the only visible trend is for underdeveloped countries to continue to be so. The development of the 20th century has led to an increasing concentration of world income, with a progressive widening of the gap between rich regions and underdeveloped countries. Underdevelopment is the manifestation of complex relationships of domination-dependency between peoples, tending to constant perpetuation under changing forms (FURTADO, 1994). As for economic development as a dynamic of capitalist accumulation within divergent models after the 2nd world war, there is the strategy of industrialization through the substitution of imports and the previous modernization of agriculture considering the promotion of exports, imports and the previous modernization of agriculture considering the promotion of exports.

At the strategic moment of industrialization, studies present the Theory of economic development in the view of the Economic Commission for Latin America and the Caribbean (ECLAC) and the main aspects of what has been going on in Brazil between cephalans and opponents; these studies highlight Prebisch's thesis, which, in turn, criticizes the theory of comparative advantage of David Ricardo who values specialization in products with lower cost advantages, that is, Latin America should, according to the theory, specialize in matters press (Souza, 1996).

Prebisch argued a downward trend in agricultural prices relative to industrial prices, thus causing a deterioration in exchange relations. His proposal saw industrialization as the only form of development based on the substitution of exports. He also indicated that the compression of superfluous consumption, the incentive to the inflow of foreign capital, the agrarian reform to increase the supply of food and the greater participation of the state in fundraising were necessary.

In general, underdevelopment is nothing more than the absence of capitalism and not its result (Mantega, 2000). However, ECLAC's ideas have been widely criticized. According to some authors, there is no empirical verification in which exchange relations would worsen against primary exporting countries. For others, poor countries with cheap labor and an abundance of natural resources would attract foreign investment, but would remain dependent on and tied to international imperialism.

As for the model of previous modernization of agriculture and the promotion of exports, it is characterized by the neoclassical / neoliberal ideas. Countries that have modernized their agriculture, such as Australia, have managed to develop from an agricultural base and sustained by the dynamism of exports. Through a Neocepalin approach, it appears that the consumer goods industries were installed on the periphery, but the capital goods industries remained at the center. This increased trade interdependence between the economies of the center and those of the periphery, but asymmetrically, since the exchange relations continued to be unfavorable for the latter. Due to the globalization process, where countries benefit from interdependence, the dependency theory has become out of fashion.

International data indicate the wide differences in income between developing countries (Souza, 1996). Average income levels in many of these countries, specifically in Latin America, are similar to the American income levels of the past century. But in other developing countries, in Asia and Africa, per capita incomes are even lower and resource exploitation is predatory. In addition, there are wide disparities in the income distribution of each country, with a small portion of the population living really well, and the majority with incomes well below the average income level. In this context, the need to consider the dimensions that are not only economic, but social and ecological presents a new normative reference, sustainable development.

Sustainable development would be a qualitative improvement that does not imply a quantitative increase greater than that acceptable for the carrying capacity, that is, the ability of the environment to regenerate raw material inputs and absorb residuais outputs (Souza, 1996).

In this previous context, studies deduce that the main challenge to be able to effectively implement sustainable development processes is the need to seek methods and ways capable of measuring and proposing changes to regulate material energy flows through economic systems. However, it is observed that the concept of sustainable development has been interpreted in the most diverse ways, always depending on the specific interests of the user. It is at this point that the present article acquires practical connotation from the realities verified in the Amazon.

The Hydroelectric potential, expressed in the expansion of the electric sector, is not proportionally translated into development, especially in sustainable development. Finally, reflecting on the interpretations of development and the environment of the energy sector, particularly the electricity sector, in the Amazon represents a timely challenge.

When analyzing the analogy and the construction of hypotheses, it is observed that in order to know how to build the object and to know the object that is constructed, it is necessary to be aware that every properly scientific object is consciously and methodically constructed, and it is necessary to know all of it to ask ourselves about the techniques of construction of the questions asked to the object. It is at this moment that we notice the opportunity to understand the terms modernity and development, as well as their intrinsic relations to create reasonable conditions for analyzing the policies of the energy sector, a strategic segment for development. When carrying out studies on policies, investments and the expansion of energy production in the Amazon and their ability to articulate development and an improvement in the living conditions of the population, they observe quite clearly that the electrification of the region was not capable of producing a socioeconomic development compatible with the huge investments made in the energy sector and with its expansion of electricity consumption. It is in this direction that all the effort built from the authors in this discussion, will serve as a basis for reflection to help understand the strategies and public actions with which different social actors act and interfere in the solution of problems pertinent to the development of the energy sector in the Amazon.

Through a rich and thought-provoking discussion, studies claim that in order to properly understand the nature of modernity we must analyze what are the sources of its dynamics. Therefore, the characteristics of modern institutions considered together may, by virtue of their dynamics, provide the basis for analyzing the planning and operationalization of the geopolitical and socioeconomic strategies of the energy sector in the Amazon. Discontinuities may, when identified, represent elements that influence the performance of policies in the region (Giddens, 1991).

Research seeks to draw conclusions from what is happening today. Investigations based on historical foundations, but failing to make predictions of possible consequences resulting from the transformation in which we live. However, the research is so rich that it can serve as a basis for studies that aim to interpret what is to come. In this way, inspired by values of responsible social participation, networks can channel the "power of flows" mentioned by the author for the implementation of regional public policies and the strengthening of democracy. This is the challenge that opens up for the energy sector, which will be able to look to the networks for a more effective instrument to promote Amazonian development.

When considering that the social contract can create a society as oppressive as Leviathan and the Enlightenment only criticizes traditional society without clarifying the mechanisms of functioning of a new society, studies collaborate expressively to the critical formation of the idea that modernity is defined by the growing separation between rationalization and subjectivation. As the author defends the idea that modernity is defined by the growing separation between rationalization and subjectification, it is observed that in the energy sector, a specific universe of my research work, the inevitable construction of absolute power is faced and repressive through its public policies, in order to lose in the adventure its internal rationality, when in the world of social actors, deprived of their identity in the name of their universal mission.

Based on a deep political concern, studies develop a simultaneous analysis of the dimensions of globalization based on a more realistic dynamic of this interpretation, which contributes significantly to the creation of conditions for a more coherent understanding (Santos, 2013). The analysis of the dimensions: economic, social, cultural and political are related to the points of evaluation of the evolution of the sustainability of the energy sector and its commitment to the proportional development. Investigations deal with energy along with dimensions, employment, equity, efficiency, among others that are interrelated to the dimensions cited as a measure of regional development (Santos, 2013). After discussing issues related to modernity and development, there is a need for proximity to what is called rationality. Investigations seek to discuss rationality as a support for the considerations of studies analyzed in previous texts (Habermas, 1987).

The highlight of the competence of sociology in the universe of this challenge represents one of his greatest observations; creates a new paradigm for sociological discussion in which it combines the lived world with the systemic conception, which constitutes a fertile field of reflection when we work on the realities of the energy sector in the Amazon and its systemic character through the energy chain, which comprises the generation, transformation, transmission, storage, distribution and consumption (Habermas, 1987). This understanding provides subsidies for analyzing the context of rationality with this follow-up.

Relevant research notes that the links between the environment, social justice and governance have become increasingly vague in some sustainability discourses, and that the structural relationships between power, conscience and the environment have been gradually obscured (Redclift, 2023). Energy merchandise, in this context, can be identified as an element to ensure a standard of quality of life, however energy comprises a field of commodity exchange relations with a view to the accumulation of capital, implying a vigorous process of exclusion. Hence the opportunity for analysis from Redclift to examine in more detail the dynamics of sustainability and to address the issue of energy sustainability in the Amazon, more specifically (Redclift, 2023).

The issue of development through dimensions that interfere with any more careful assessment of the notion of development, is what addresses studies that find that current development patterns, said to be sustainable, break the relations between social systems and ecosystems, instead of ensuring that the use of natural resources by communities meets their needs at a level of comfort assessed as satisfactory for these communities, the author demonstrates that he has a rich and coherent analysis of the theme (Banerjee, 2003).

Understanding the concept of sustainable development, as well as discussing the implications of this concept with contemporary analyzes of biodiversity, biotechnology and intellectual property rights naturally stimulate the formulation of sustainable development alternatives and the implications for theory and practice management of natural resources. It is at this moment that I relate to water resources, which promote energy through hydroelectricity, through the possibility of better articulating the strategies developed by the public policies of the energy sector in the Amazon from the capitalist notions of managerial efficiency.

PUBLIC ELECTRICITY MANAGEMENT

Public planning is technically evidenced through the events of the budget cycle, which, in turn, is composed of the phases in which public budget preparation and execution activities occur (Giacomoni, 2010). However, the debate about public planning has frequently promoted mentions, especially, of the ability of managers to achieve qualitative results in the use of public resources applied to the territory; among them, the influence of group ideologies that interfere with more decision-making power, through correlations of forces with various ramifications, such as energy (Mafra & Silva, 2004; Schultz, 2016; Dagnino *et al.*, 2016).

In the environment of the Brazilian energy sector, planning is developed through public policies, which generally intend to demonstrate that investments aim at economic growth and improvement of the population's living conditions. In this process, there are strategic aspects ranging from the choice of energy generation sources to the reflections of the use of this energy in different sectors of a country's economy (Borges 2012; Cornescu & Adam, 2014). The projection of the composition of these available energy sources, which should be directed to meeting the energy demands in a given state, region or country, is called the energy matrix.

The energy matrix is the description of all the generation and consumption of energy, of a certain spatial cut, broken down in terms of production sources and consumption sectors for a future situation; thus, when all the generation and consumption of a country or region is described for a current situation, it is called energy balance (Borges & Zouain, 2010; Reis *et al.*, 2012). According to EPE (2020), the national energy matrix is currently prepared by the Energy Research Company. The electrical matrix, in turn, is inserted in the energy matrix and represents the disposition of the different forms, specifically, of electricity, made available to the productive processes in a determined spatial context, involving its sources of generation and use (Tolmasquim *et al.*, 2007; Reis *et al.*, 2012).

The observance of this arrangement of sources for the generation of electricity assumes a strategic role insofar as the projections verified in a given electrical matrix, value the facilitation of forms of access to the population. Electric energy sources comprise essential inputs for sustainable development (Goldemberg & Moreira, 2005) and the understanding of this normative reference is essential for the construction of an electrical matrix.

The conceptualization of sustainable development assumes a new world order, which results in a redistribution of powers that ignores the correlations of forces that are active in the world market, and the interests of industrialized nations in maintaining a position of advantage in the international panorama (Borges *et al.*, 2017; Silva *et al.*, 2018). However, the definition of this normative reference is steeped in contradictions, insofar as the difficulty lies in the fact that economic interests are not submissive to social and environmental interests. Sustainable development aims to promote sustainability.

Sustainability is linked to an activity that can be maintained for an indefinite period of time, so as not to reach its exhaustion, despite the unforeseen events that may occur during this period, based on relatively consistent economic, social and environmental bases (Glavic & Lukman, 2007; Marzall, 1999; Bursztyn, 2008; Borges, 2012). According to Camargo *et al.* (2004), Bursztyn (2008) and Borges (2015), sustainability would then be defined as the ability to sustain socioeconomic and environmental conditions that promote the fulfillment of human needs in a balanced way and that occurs through decision-making.

According to Robbins (2010), decision making comprises an occurrence in reaction to a problem and a problem exists when there is a discrepancy between the current state of affairs and its desirable state. In the theoretical framework, decision-making comprises the process of deciding on something and involves selecting an action option from two or more possible alternatives; and in this process there would be at least three components that apply to the public environment and that need to be verified by the observer who wants to carry out investigations on the decision-making process in this area. They are: technology; rules and norms; and decision-making style (Silva, 2013). Regardless of its components, decision-making in public planning shows an environment of frequent slowness due to excessive bureaucracy (Pacheco & Mattos, 2014). In the decision-making process, it is important to highlight that it constitutes an error-prone activity, as it will be affected by the personal characteristics and perception of the decision-maker (Maximiano 2009; Robbins, 2010).

In this perspective, more impulsive decision-making and less information processing threatens decision-making processes. In the modern context, decision-making carries cognitive biases and these biases interfere with our decision-making in ways that move away from rational objectivity (Certo, 2005, Colson, 2019). It is precisely there that lies the advent of a new phase of evolution in the decision-making environment, Artificial Intelligence. The term A.I. it is a sub-area of computer science and is used to designate the set of computational techniques, devices and algorithms, in addition to statistical and mathematical methods capable of reproducing some of the human cognitive abilities; in other words, it is the science and engineering of making intelligent machines, especially intelligent computer programs (Toffoli, 2018; McCarthy, 2018).

MACHINE LEARNING

AI comprises an area that is related to the ability of computers to perform tasks, in order to contribute to the construction of intelligent entities through computational algorithms (Norvig & Russell, 2021). In this sense, the A.I. it would consist of the artificial reproduction of the ability to obtain and apply different skills and knowledge to solve a given problem, solving it, reasoning and learning from situations (Hartmann & Silva, 2019). In other words, the A.I. It is a set of instructions and rules that form the algorithm, used in series, to process information and solve problems, with its own method and speed (Corvalán, 2018). The A.I. is already a reality and a great ally of public planning, thanks to the agility and time savings provided by the verification and crossing of data, where it creates possibilities beyond human capacity, offering public bodies elements that could go unnoticed in the analyzes commonly carried out by civil servants (Brega, 2012; Desordi & Bona, 2020).

According to the Association for the Advancement of Artificial Intelligence (AAAI), an entity recognized as a reference association, AI has subareas according to their applicability. AAAI has established a division of AI into the following subareas: Research; Machine Learning; Automated Planning; Knowledge Representation; Data Mining and Big Data; Reasoning (Probabilistic or not); Natural Language Processing; Robotics; Agent and Multi-Agent System and Applications (AAAI, 2021). In this perspective, it should be noted that this article especially addresses the Machine Learning subarea, in that it seeks to carry out an inventory of which data would be important to identify relevant patterns for decision-making in the face of electrical matrices.

Machine learning comprises a data analysis method that automates the construction of analytical models, based on the idea that systems can learn from data, identify patterns and make decisions with minimal human intervention (Colson, 2019; Desordi & Bona, 2020). In this dynamic of learning from data and identifying patterns for decision-making, the strategic opportunity of applying AI to public planning of sustainable energy matrices stands out, as analytical models capable of accurately assessing potentialities, limitations and impacts of the individual or combined use of electric energy sources in certain regions and based on the productive profile of each sector of economic activity. The possibility of effectively establishing a clear and objective link between the public planning of electrical matrices and AI lies in the construction of Algorithms. Algorithm comprises a finite sequence of executable actions that aim to obtain a solution to a given problem (Dasgupta et al., 2010). In this perspective, the Algorithms applied to AI are constituted as a sequential set of rules or operations that, applied to a number of data, enable the resolution of problems.

STRUCTURE OF THE DECISION-MAKING MODEL

This investigation is categorized in terms of ends and means, according to Vergara's grouping (2016). As for the purposes, it is considered exploratory, insofar as it involves a survey of subsidies that propose to analyze realities that stimulate the understanding of the connections between A.I. and the planning of sustainable electrical matrices. And as for its means, it is considered bibliographic and documental, insofar as it uses a survey of materials and documents from bodies linked to the Brazilian electricity sector.

The methodology was strategically divided into three tasks: data collection, data organization and data analysis.

DATA COLLECTION PROCEDURE

Initially, it is highlighted that the variables used in the construction of the indicators were: amount of electricity consumed; Gross Domestic Product; amount invested in electricity; average electricity tariff; average income of the worker; number of jobs generated; energy efficiency and amount of polluting gas emissions. The period of data collection with the variables was between 2008 and 2018. The spatial collection of this data was the State of Pará, chosen for its characteristic of being an exporter of electricity and for being a place of socio-environmental impacts resulting from the process of generating energy from a water source. The sources used in the research for the elaboration of the indicators were: National Energy Balance (BEN); Useful Energy Balance (BEU); National Household Survey (PNAD); National Electric Energy Agency (Aneel), Secretary of Planning, Budget and Finance of the State of Pará (Sepof); Brazilian Institute of Geography and Statistics (IBGE); Interunion Department of Statistics and Socioeconomic Studies (Dieese), General Register of Employed and Unemployed People (Caged); National Institute of Energy Efficiency (INEE);

Ministry of Mines and Energy (MME); National System Operator (ONS); and Energy Equatorial Group (EEG).

DATA ORGANIZATION PROCEDURE

In this study, two analysis structures were constructed. The first refers to the use of electrical energy sources, based on the economic, social, environmental and technological dimensions. The second structure of analysis, pertinent to the use of sectoral indicators of energy sustainability, supported by linear correlations of an economic, social, environmental and political nature. In the process of elaborating the structure of analysis pertinent to the use of electric energy sources, supported by the economic, social, environmental and technological dimensions, emphasis was placed on the identification of potentialities and limitations in the use of electricity sources: water, biomass, solar, wind and nuclear. The intention was to provide a critical examination regarding the insertion possibilities and proportion of these sources in the electrical matrix. The dimensions used in this analysis were economic, social, environmental and technological, as they better characterize the environment of sustainable development. It is important to point out that this critical examination of electrical energy sources must respect the locational specificities, as each region presents strategic potential and substantial limitations regarding the production and use of the electrical input. In the elaboration of the sectorial analysis structure of energy sustainability indicators, supported by linear correlations, the correlation sought as a result a coefficient that quantified the degree of correlation called Pearson's coefficient (p) (Chen & Popovic, 2002).

$$r = \frac{\sum_{i=1}^{n}(x_i - \bar{x})(y_i - \bar{y})}{\sqrt{\sum_{i=1}^{n}(x_i - \bar{x})^2} \cdot \sqrt{\sum_{i=1}^{n}(y_i - \bar{y})^2}},$$

Where: x1 , x2 , ..., xn and y1 , y2 , ..., yn comprise the measured values of both variables. And the following equations are the arithmetic means of these variables:

$$\bar{x} = \frac{1}{n} \cdot \sum_{i=1}^{n} x_i \qquad e \qquad \bar{y} = \frac{1}{n} \cdot \sum_{i=1}^{n} y_i$$

In this study, the linear correlations observed in each of the dimensions, across sectors, were described and analyzed with regard to their importance, representativeness and unit of measurement used. The software used for calculation was the Statistical Package for the Social Sciences (SPSS). At a later moment, the variables were organized according to the dimensions: economic, social, environmental and political, which built the energy sustainability indicators, and from each sector of activity, which made up the energy sustainability indexes. In calculating the indicators, we proceeded from a weighted average composed of the result of the calculation of the composite variables. In calculating the composite variables, the calculation adopted two variables: the first referring to development, and the other referring to the energy environment. The results of the correlations, which indicated the use of certain relationships of variables to the detriment of others, were due to the specificities of the data referring to the State of Pará, between 2010 and 2019, used here to merely operate the calculation and consequent indication of the relationships of variables across sectors of economic activity. Thus, depending on the state or region and based on their respective locational specificities, the relationships of variables indicated by the software used can naturally be different.

DATA ANALYSIS PROCEDURES

The analysis structures of electricity generation sources and energy sustainability indicators, built in the study, were examined with the intention of raising subsidies in the construction of decision-making standards along with the process of elaboration of the electric matrix. The purpose of this examination is to verify the possibility that the A.I. build analytical models capable of evaluating, with precision, potentialities, limitations and impacts of the individual or combined use of electric energy sources in certain regions and based on the productive profile of each sector of economic activity. In the practical field, the possibility of effectively demonstrating the connections between public planning of electrical matrices and AI lies in the construction of Algorithms. In this sense, Algorithms applied to AI would be guidelines to be learned and followed by a machine.

In this study, these guidelines that would feed the Algorithms would be assigned precisely from the result of the analysis of the structure of electricity generation sources and the structure of energy sustainability indicators. In reality, Algorithms would just be a mathematical way of demonstrating a structured process for executing a task. In other words, they would be principles and flows of sequential analysis that guide the decision-making process in the electricity sector. In this methodological strategy, it was intended to present only a first approximation to the effort to build Algorithms at the service of AI, through the basic phases of an Algorithm's performance: input, processing and output, in the context of sustainable electrical matrices.

ANALYSIS OF THE DECISION-MAKING MODEL

This section is divided into two parts, namely: analysis of electricity generation and analysis of energy indicators. Analyzes should start from a complete and integrated data structure that needs to be classified and grouped in order to synergistically produce mental captures and predictions; and it is with the intention of favoring a coherent analysis that this didactic division was adopted.

ANALYSIS OF ELECTRICITY GENERATION

This subsection addresses water, biomass, solar, wind and nuclear sources, which will precede the presentation of an analysis structure of electrical energy sources, by dimension.

a) Water source: based on simple fundamentals - Turbines transform potential energy from reservoirs or water currents into electrical energy. Thus, hydroelectric plants are characterized as renewable energy. However, this source causes serious and extensive impacts on the hydrological cycle and changes in the environment in general. The results record the disappearance of species of fauna and flora, loss of quality of life for the affected populations and threats to the existence of various social groups. The emission of greenhouse gases represents another serious problem caused by large hydroelectric plants, a panorama that does not portray socioeconomic and environmental conditions that promote the fulfillment of human needs in a balanced way (Camargo *et al.*, 2004; Bursztyn, 2008; Borges, 2015).

b) Biomass: is a type of matter that feeds steam power plants for electrical generation from a process of burning elements accumulated in a given ecosystem. Among the most used materials are sugarcane bagasse and woody materials. The burning of biomass causes the release of carbon dioxide into the atmosphere, however, this compound was previously absorbed by the plants that gave rise to the fuel, which provides a zero balance of CO_2 emissions. It is also important to mention that these materials must be close to thermoelectric plants or on strategic routes with easy access, otherwise they may represent disadvantages. In this perspective, the reflections resulting from the use of a certain energy source must be critically observed (Bermann, 2003; Borges 2012; Cornescu & Adam, 2014).

c) Sun: consists of the use of thermal and light energy captured by solar panels, consisting of photoelectric or photovoltaic cells. This type of energy source is considered clean, renewable and inexhaustible. Solar energy is the one that releases the least CO_2 into the atmosphere, in addition to generating potential for job creation in the solar production chain. The main disadvantages of the solar source are: the high cost of implanting thermosolar panels, which are too expensive to enable the production of electricity on a large scale, and its irregularity in the form of uniform distribution, which requires large collection areas and storage systems. . In this analysis bias, according to Borges (2012); Cornescu & Adam (2014), the reflections resulting from the use of a certain source of energy in the environment must be critically observed.

d) Wind power: energy derived from a technology that uses the power of wind which, in turn, operates turbines connected to electricity networks. This type of energy source has the advantages of renewable nature, low cost of externalities, does not burn fossil fuels and does not emit polluting gases that cause the greenhouse effect. Among the disadvantages are: the alteration of the landscape when implementing its infrastructure, composed of propellers and towers, the emission of low frequency noise, occasional interference in television sets, the threat to the migratory routes of birds due to the use of large rowed propellers and the unproductiveness of this source in some regions due to the inconstancy of winds, their low intensity and waste of energy in the event of heavy rains.

e) Nuclear: it takes place from a thermal base, where the heat produced in the fission to generate water vapor that moves the electricity generation turbines. The advantage of this source lies in its technology, capable of reducing gas emissions in the production of electricity and the climate impacts on the planet caused by the generation of electricity. However, it is relevant to alert defenders of nuclear technology as not emitting greenhouse gases, that if the calculations of the complete process of this type of energy are incorporated, including: uranium mining, transport, uranium enrichment, the subsequent dismantling of the central and the processing and confinement of radioactive waste, this source has disadvantages.

As a methodology that automates the analysis of information, according to Colson (2019), Desordi and Bona (2020), machine learning can favor the development of structures for analyzing electrical energy sources, by dimension, (Table 1) that allow an examination more accurate and ideal for choosing the arrangement of electrical matrix components.

Table 1 - Dimensional Analysis of Electricity Generation.

BASE	DIMENSION	GENERAL INFORMATION
WATER	ECONOMIC	Reservoir construction costs and after construction.
WATER	SOCIAL	Generation of jobs in the construction of reservoirs and after construction.
WATER	ENVIRONMENTAL	Emissions of polluting gases.
WATER	TECHNOLOGICAL	Energy density.
BIOMASS	ECONOMIC	Construction costs of a small plant and estimated average cost.
BIOMASS	SOCIAL	Job creation.
BIOMASS	ENVIRONMENTAL	Emissions of polluting gases and capacity for devastation.
BIOMASS	TECHNOLOGICAL	Generation capacity and capacity corresponding to generation.
SUN	ECONOMIC	Installation costs of photovoltaic system and rate of return on investment.
SUN	SOCIAL	Job generation.
SUN	ENVIRONMENTAL	Emissions of polluting gases in the construction of the plant and in its operation.
SUN	TECHNOLOGICAL	Solar radiation potential and infrastructure characteristics.
WIND	ECONOMIC	Cost of installing a wind farm and return on investment.
WIND	SOCIAL	Job generation.
WIND	ENVIRONMENTAL	Emissions of polluting gases in the construction of the plant and in its operation.
WIND	TECHNOLOGICAL	Wind density and characteristics.
NUCLEAR	ECONOMIC	Cost of installing a wind farm and return on investment.
NUCLEAR	SOCIAL	Job generation.
NUCLEAR	ENVIRONMENTAL	Installation costs and return on investment.
NUCLEAR	TECHNOLOGICAL	Energy intensity.

Source: Prepared by the author (2024).

Decision-making based on the analysis of the use of electrical energy sources, by dimension, is inserted in environments with multiple information (endowed with inaccuracies) and in environments where knowledge is extracted from numerous databases. The high risk in this process may represent high environmental costs or severe socioeconomic damage to the affected populations, due to the results of this decision-making process. In this sense, the A.I. adds: the reduction in the analysis time of the most indicated sources; the solution of problems verified with the demands of society, for the electrical input, whether operational or managerial; and the reduction of rework and failures, frequent in electric energy generation projects observed in Brazil.

ANALYSIS OF ENERGY INDICATORS

The construction process of an electrical matrix must also consider aspects related to economic sectors, since each sector of economic activity reflects investments in electrical energy based on certain peculiarities. These peculiarities can be organized into economic, social, environmental and political dimensions. Below, through Tables 2, 3, 4 and 5, the structures of energy sustainability indicators for each sector of activity are presented, based on linear correlations and from data referring to the State of Pará, during the period 2010 and 2019. Data referring to the state of Pará were used only with the intention of considering its locational specificities.

Table 2 - Analysis of electricity sustainability indicators (Agricultural Sector).

AGRICULTURAL SECTOR	ECONOMIC	1) Relationship between the value of the Gross Domestic Product in the agricultural sector and the amount of GWh consumed in the sector. 2) Ratio between the amount invested by the distributor from Pará in electricity in the State and the value of the Gross Domestic Product, per unit of consumption, in the agricultural sector. 3) Relationship between the average electricity tariff charged per kWh in the agricultural sector and the Gross Domestic Product, per unit of consumption, in this sector. 4) 4) Ratio between the amount invested by the distributor from Pará in electricity in the State and no. of consumption units in the sector.
	SOCIAL	1) Relationship between the amount of GWh consumed in the agricultural sector and the average income of Pará workers. 2) Relationship between the amount of GWh consumed in the agricultural sector and the Gini coefficient recorded in the State of Pará.
	ENVIRONMENTAL	1) Relationship between the amount of GWh consumed in the agricultural sector and the energy yield verified in this sector. 2) Relationship between the amount of GWh consumed in the agricultural sector and the accumulated emission of methane gas (CH_4) and carbon gas (CO_2) derived from hydroelectric plants in the state of Pará.
	POLITICAL	1) Relationship between the average electricity tariff charged per kWh in the agricultural sector and the equivalent frequency of interruption per consumer unit in all sectors of the State. 2) Relationship between the amount of GWh consumed in the agricultural sector and the equivalent frequency of interruption per consumer unit in all sectors of the State.

Source: Prepared by the author (2024).

The structure of indicators shown in Table 3, like the other structures built, was created based on different skills and knowledge, which in a combined way seeks to solve problems related to the use of electrical input, in each sector of activity, in favor of improvement of the standard of living of the population of Pará. In this sense, the A.I. it would consist of the artificial reproduction of this ability to obtain and apply these skills and knowledge to solve a given problem, solving it, reasoning and learning from situations (Hartmann & Silva, 2019). These indicator structures can help build the electrical matrix based on strategies to promote sustainable development.

Table 3 - Analysis of electricity sustainability indicators (Industrial Sector).

INDUSTRIAL SECTOR		
ECONOMIC	1)	Relationship between the value of the Gross Domestic Product in the industrial sector and the amount of GWh consumed in the sector.
	2)	Relationship between the value of the Gross Domestic Product in the industrial sector and the number of consumption units in the sector.
	3)	Relationship between the average electricity tariff charged per kWh in the industrial sector and the Gross Domestic Product in this sector
	4)	4) Relationship between the amount invested by the distributor from Pará in electricity in the State and the number of consumption units in the sector.
SOCIAL	1)	Ratio between the amount invested by the distributor from Pará in electricity in the State and the number of jobs generated in the industrial sector.
	2)	Relationship between the amount of kW consumed in the industrial sector and the Gini coefficient recorded in Pará.
	3)	3) Relationship between the number of consumption units in the industrial sector and the recorded Gini coefficient.
ENVIRONMENTAL	1)	Relationship between the amount of GWh consumed in the industrial sector and the energy yield verified in this sector.
	2)	2) Relationship between the amount of GWh consumed in the industrial sector and the accumulated emission of methane gas (CO_2) and carbon gas (CO_2) derived from hydroelectric plants in the state of Pará.
POLITICAL	1)	Relationship between the average electricity tariff charged per kWh in the industrial sector and the equivalent frequency of interruption per consumer unit in all sectors of the State.
	2)	2) Relationship between the number of consumer units in the industrial sector and the equivalent interruption frequency per consumer unit in all sectors.

Source: Prepared by the author (2024).

In Table 4, there is an effort whose result is based on a statistical tool that seeks to help the decision maker in the process of building the electrical matrix. However, as noted by Toffoli (2018) and McCarthy (2018), what A.I. may add, is something beyond statistical and mathematical methods, that is, the development of an ability to reproduce some of the human cognitive abilities.

AI, according to the author Certo (2005), Maximiano (2009) and Robbins (2010), will enable greater security, trust and control over the interactions necessary for a complex and multidisciplinary process such as the construction of an electrical matrix, since that the traditional decision-making process is prone to errors.

The A.I. aspects that may negatively impact the process of building energy matrices should also be the focus of attention. Moral and ethical questions, as a result of the use of an innovative technology, must be worked on carefully. The dimensions analyzed in an energy matrix are relevant to the strategic areas of sustainability, however, it involves the reality of people and possibilities for improving or not the standard of living. In short, the intention is that the A.I. can be used in a prudent and transparent manner as a technology capable of carrying out an advanced and strategic reading of society's energy demands.

Table 4 - Analysis of electricity sustainability indicators (Commercial sector).

COMMERCIAL SECTOR		
	ECONOMIC	1) Relationship between GDP in the sector and the amount of KW consumed in this sector. 2) Relationship between the amount of GW consumed and the amount invested in electricity in all sectors. 3) Relationship between the variation in the electricity tariff and the amount invested in electricity in all sectors
	SOCIAL	1) Relationship between the balance of formal jobs in the sector and the amount invested in electricity. 2) 2) Relationship between average income and the amount of GW consumed.
	ENVIRONMENTAL	1) Relation between the variation of energy yield in the sector/Amount of GW consumed. 2) 2) Relationship between the variation in the emission of polluting gases derived from electricity generation and the amount of GW consumed.
	POLITICAL	1) Ratio between the variation in the equivalent frequency of interruption per consumer unit and the variation in the tariff charged for electricity 2) 2) Relationship between the variation in the duration of interruptions per consumer unit and the variation in the tariff charged for electricity in the sector.

Source: Prepared by the author (2024).

As presented in the methodology of this investigation, the possibility of effectively establishing a direct link between the public planning of electrical matrices and AI is linked to the construction of Algorithms. The Algorithms applied to AI would be principles and rules (guidelines) to be followed by a machine, that is, a structured process for the execution of a task that proposes to guide decision-making in the construction of sustainable electrical matrices. Next, through Chart 6, the foundations for the development of algorithms for Artificial Intelligence in favor of sustainable public management of electric energy are presented.

Table 5 - Foundations for the development of algorithms for AI in favor of sustainable public management of electric energy.

STAGE	INPUT STAGE	PROCESSING STAGE	OUTPUT STAGE
GUIDE	INITIAL DATA, THAT IS, POSSIBLE INPUT DATA VALUES OF AN ALGORITHM.	RELATIONS THAT MUST BE SATISFIED TO TRANSFORM INPUT DATA INTO ACCEPTABLE OUTPUT.	RESULTS, THAT IS, DISPLAY OF THE CALCULATION FROM THE INPUT AND PROCESSING PHASES.
INSTRUCTIONS TO ARTIFICIAL INTELLIGENCE	Amount of electricity generated in GW; Amount invested for the generation of electricity; Amount of electricity consumed; Accumulated emission of carbon dioxide gas (CO_2) derived from hydroelectric plants; Energy yield; Electricity generation capacity in Kw; No. of jobs generated from the source of generation; Costs used in generation projects.	Ratio between the amount of electricity generated in GW and the amount invested in generating electricity; Relation between the accumulated emission of carbon dioxide gas (CO_2) derived from hydroelectric plants and the amount of GWh consumed; Relationship between energy yield and amount of GWh consumed; Costs used in generation projects per Kw; Relationship between number of jobs generated from the source of generation and the Amount invested for the generation of electricity through this source.	Ability to meet energy demand; Economic externalities; Social externalities; Environmental externalities; Energy-material flow.

Source: Prepared by the author (2024).

The results where the generation of electricity and its combined disposition structure in favor of a population, is presented from sustainable bases, must be codified in a programming language. In the coding process, programming language tools must be used. The Java Development Kit (JDK), which is a utility suite that makes it possible to build software systems, and the NetBeans Integrated Development Environment (IDE), which is a program for software development.

The technique for building an Algorithm indicated for the electrical matrix environment would be the so-called Narrative Description, which defines what needs to be done and how to do it, that is, it identifies the steps to be followed to achieve sustainable solutions in the electrical matrix environment. The steps related to the use of programming language tools and the complete elaboration of the Algorithms are not an object of observation here.

ADVANCES IN THE PROPOSED DECISION-MAKING MODEL

The possibility that the A.I. building analytical models capable of guiding decision-making in the public planning of sustainable electricity matrices is concrete. In this perspective, this investigation constitutes an original contribution insofar as it discusses the possibility of a connection between artificial intelligence and sustainable public management of electricity in Brazil. The analysis structures elaborated in this investigation, despite being synthetic, aggregate reasonable conditions for evaluating the potentialities, limitations and impacts of the individual or combined use of electric energy sources, in certain regions, and based on the productive profile of each sector of activity.

In this sense, it is up to machine learning, processing and methodological elaboration, based on algorithms, for a precise and transparent examination of information that automates the construction of decision analysis models, supported by the idea that computational systems can learn from data, identify standards, strategically consider locational specifics, and make decisions with minimal human intervention. The discussion about these connections helps to raise subsidies for machine learning to process and develop methodologies, based on algorithms, that automate the construction of decision analysis models in the elaboration of sustainable electrical matrices. The investigation inferred that the possibility of effectively establishing a direct link between the public planning of electrical matrices and AI is linked to the construction of Algorithms. In this perspective, the investigation presented, as an effort to raise subsidies, preliminary stages of construction of Algorithms for AI applied to the decision-making process in the planning of sustainable electrical matrices. This effort constituted a structured proposal of guidelines for machine learning that proposes to contribute to the construction of Algorithms aimed at guiding decision-making in the construction of sustainable electrical matrices.

The research concluded, in this dynamic of analysis, that artificial intelligence can guide decisions in the planning of electrical matrices, as long as they are based on analysis structures focused on the strategic use of electricity sources and on the use of sectoral and multidimensional indicators. These analysis frameworks are able to feed computational analytical models that learn from data, identify patterns and make intelligent decisions with minimal human intervention. Future research on this topic should follow the path of improving the process of building Algorithms for AI applied to the decision-making process, in order to still observe the environment of the sectors of economic activity. The use of energy input in the intricacies of each sector of economic activity presents different reflections on the development process, which represent different intensities in terms of jobs generated, energy efficiency, contribution to GDP, favoring the deconcentration of income, among other relevant and strategic variables in the development process.

BIBLIOGRAPHIC REFERENCES

Araújo, V. S.; Zullo, B.A.; Torres, M. Big data, algorithms and artificial intelligence in public administration: reflections for its use in a democratic environment. *Journal of Administrative and Constitutional Law.* v. 20, n. 80, 2020.

Association for the Advancement of Artificial Intelligence [AAAI]. *In: the official publication of the AAAI.* Available at: https://www.aaai.org/home.html Accessed: August 15, 2021.

Banerjee, S. B. Who sustains whose development? Sustainable development and the reinvention of nature. *In: Fernandes, M. and Guerra, L. (Org) Counterdiscourse on sustainable development.* Belém: Unamaz, 2003.

Brega, J. F. F. *Electronic government and administrative law.* 2012. 336 f. (Doctoral Thesis) - Faculty of Law, University of São Paulo, São Paulo, 2012. http://www.teses.usp.br/teses/disponiveis/2/2134/tde-06062013-154559/publico/TESE_FINAL_Jose_Fernando_Ferreira_Brega. pdf. Accessed on 12 nov. 2020.

Borges, F. Q.; Zouain, D. M. The electrical matrix in the state of Pará and its position in promoting sustainable development. *Planning and public policies,* n. 35, jul./dec, 2010.

Borges, F. Q. Public administration in the electricity sector: sustainability indicators in the residential environment in the state of Pará (2001-10). *Rev. Adm. Public [online].* vol.46, n.3, p.737-751, 2012.

Borges, F. Q. Institutional sustainability in the Brazilian electricity sector. *Pretext Magazine.* vol. 16, no.1, jan./mar. 2015.

Borges, F. Q.; Rodrigues, I.M.; Oliveira, A. S. P. Paradox of electricity in the state of Pará: a study of the factors that contribute to high residential tariffs (2005-2014). *Observatorio de la Economía Latinoamericana.* Intercontinental Academic Services. Malaga: Issue 231, May, 2017.

Bursztyn, M. State and environment in Brazil. *In: Thinking about sustainable development.* São Paulo, Brasiliense, 2008.

Camargo, A. S. G.; Ugaya, C.M.L.; Agudelo, L. P. P. Proposal for defining sustainability indicators for electricity generation. *Education and Technology Magazine*, Rio de Janeiro: Cefet/PR/MG/RJ, 2004.

Cardoso, F. H. *Ideas and their place*. Petrópolis: Voices, 1993.

Colson, E. What AI-Driven Decision Making Looks Like. *Harvard Business Review*. Jul, 8, 2019.

Chen, P.Y.; Popovic, P.M. *Correlation*. London: Sage, 2002.

Corvalán, J. G. Digital and Intelligent Public Administration: transformations in the Era of Artificial Intelligence. *A&C Journal of Administrative and Constitutional Law*, Belo Horizonte, year 18, n. 72, p. 81-82, jan./mar, 2018.

Dasgupta, S.; Papadimitriou, C.; Vazirani, U. *Algorithms*. Porto Alegre: AMGH, 2010.

Desordi, D.; Bona, C. D. Artificial intelligence and efficiency in public administration. *Law Magazine*. Lush. v.12 n.2, 2020.

Dreyfus, S. E. & Dreyfus, H. L. *A five-stage model of the mental activities involved in directed skill acquisition*. Berkeley, University of California: Operations Research Center, 1980.

Energy Research Company [EPE]. Available at: www.epe.gov.br/Paginas/default.aspx Accessed on: January 20, 2020.

Energy Information Administration [EIA]. *International energy outlook*. Washington: EIA, 2018.

Figueiredo, C. R. B. & Cabral, F. G. Artificial intelligence: machine learning in Public Administration. *International Journal of Digital Law*. v. 1 n. 1, 2020.

Furtado, C. *Dialectic of development*. Rio de Janeiro: Culture Fund, 1994.

Glavic, P.; Lukman, R. Review of sustainability terms and their definitions. *Journal of Cleaner Production*, v.15, p.1875-1885, 2007.

Giacomoni, J. *Public Budget*. São Paulo: Atlas, 2010.

Giddens, A. *The consequences of Modernity*. São Paulo: Unesp, 1991.

Habermas, J.*Theory of communicative action*. Madrid: Taurus, 1987.

Hartmann, F.P.; Silva, R. Z. M. da. *Artificial intelligence and law*. Curitiba: Alterity, 2019.

Mafra, F. & Silva, J. A. *Planning and management of the territory*. Porto: Portuguese innovation Society, 2004.

Mantega, G. *The Brazilian political economy*. Petrópolis: Voices, 2000.

Marques, K. V. S. *The administrative act and artificial intelligence*: an approach to the limits and possibilities of using artificial intelligence in the context of public administration. (Monograph). Federal University of Rio Grande do Norte, Center for Applied Social Sciences, Graduate Program in Law. Christmas: UFRGN, 2020.

Marzall, K. *Sustainability indicators for agroecosystems*. Porto Alegre: UFRGS, 1999.

Maximiano, A. C. A. *Introduction to Administration*. São Paulo: Atlas, 2009.

McCarthy, J. *What is artificial intelligence?* Stanford. http://jmc.stanford.edu/articles/whatisai/whatisai.pdf. Accessed on: 28 dec. 2022.

Norvig, P.; Russell, S. J. *Artificial Intelligence*: a modern approach. United Kingdom: Pearson, 2021.

Parsons, T. Evolutionary Universals in Society. *In: from modernization to globalization.* s/l: s/e, 1964.

Power Plants in Northern Brazil. *Report on the use of alternative sources of electricity*. Brasília, 2010.

Power Stations of Pará [Celpa]. *Management report*: fiscal year 2010-2019. Bethlehem, 2020.

Right. S.C. Decision making. *In: Right. S.C. Modern Management*. São Paulo: Pearson, 2005.

Redclift, M. The new speeches of Sustainability. *In: Fernandes, M. and Guerra, L. (Org) Counter-discourse on sustainable development.* Belém: Unamaz, 2003.

Reis, L.B.; Fadigas, E.A.A.; Carvalho, C. E. *Energy, natural resources and the practice of sustainable development.* São Paulo: Manole, 2012.

Robbins, S.; Judge, T.; Sobral, F. *Organizational behavior.* theory and practice in the Brazilian context. São Paulo: Pearson, 2010.

Rostow, W. (1961). *Stages of economic development, a non-communist manifesto*. Rio de Janeiro: Zahar, 1961

Santos, B. de S. *Globalization; fatality or utopia?* São Paulo: s/e, 2013.

Santos, M. A. da S. *Artificial Intelligence*. S/l: Brasil Escola, 2021.

Silva, J. F. B. A.; Rebouças, S. M. D. P.; Abreu, M. C. S. de; Ribeiro, M. da C. R. Construction of a sustainable development index and spatial analysis of inequalities in municipalities in Ceará. *Rev. Adm. Public [online]*. Rio de Janeiro 52(1): p.149-168, jan.-feb., 2018.

Souza, N. de. *Economic development*. São Paulo: Atlas, 1996. (Souza, 1996).

Toffoli, D. In: Fernandes, R. V. De C.; Carvalho, A. G. P. de (Coord.). Legal technology & digital law: *In: II International Congress of Law, Government and Technology*. Belo Horizonte: Forum, 2018.

Tolmasquim M.T.; Guerreiro, A. & Gorini, R. Prospective view of the Brazilian energy matrix: energizing the country's sustainable development. *Brazilian Energy Magazine*, vol. 13, n.1. Rio de Janeiro: SBPA, 2007.

Vergara, S. C. *Management Research Projects and Reports*. São Paulo: Atlas, 2016.

Valle, V. L. do Artificial intelligence incorporated into Public Administration: myths and theoretical challenges. *A&C - R. de Dir. Adm. Const.* Belo Horizonte, year 20, n. 81, p.179-200, jul./sept., 2020.

www.ingramcontent.com/pod-product-compliance
Lightning Source LLC
LaVergne TN
LVHW080214200726
843506LV00006B/353